Jan Ślusarek

# Effects Of Cement Materials Hardening Processes In Massive Structures

AF168226

Jan Ślusarek

# Effects Of Cement Materials Hardening Processes In Massive Structures

LAP LAMBERT Academic Publishing

**Impressum / Imprint**

Bibliografische Information der Deutschen Nationalbibliothek: Die Deutsche Nationalbibliothek verzeichnet diese Publikation in der Deutschen Nationalbibliografie; detaillierte bibliografische Daten sind im Internet über http://dnb.d-nb.de abrufbar.

Alle in diesem Buch genannten Marken und Produktnamen unterliegen warenzeichen-, marken- oder patentrechtlichem Schutz bzw. sind Warenzeichen oder eingetragene Warenzeichen der jeweiligen Inhaber. Die Wiedergabe von Marken, Produktnamen, Gebrauchsnamen, Handelsnamen, Warenbezeichnungen u.s.w. in diesem Werk berechtigt auch ohne besondere Kennzeichnung nicht zu der Annahme, dass solche Namen im Sinne der Warenzeichen- und Markenschutzgesetzgebung als frei zu betrachten wären und daher von jedermann benutzt werden dürften.

Bibliographic information published by the Deutsche Nationalbibliothek: The Deutsche Nationalbibliothek lists this publication in the Deutsche Nationalbibliografie; detailed bibliographic data are available in the Internet at http://dnb.d-nb.de.

Any brand names and product names mentioned in this book are subject to trademark, brand or patent protection and are trademarks or registered trademarks of their respective holders. The use of brand names, product names, common names, trade names, product descriptions etc. even without a particular marking in this works is in no way to be construed to mean that such names may be regarded as unrestricted in respect of trademark and brand protection legislation and could thus be used by anyone.

Coverbild / Cover image: www.ingimage.com

Verlag / Publisher:
LAP LAMBERT Academic Publishing
ist ein Imprint der / is a trademark of
OmniScriptum GmbH & Co. KG
Heinrich-Böcking-Str. 6-8, 66121 Saarbrücken, Deutschland / Germany
Email: info@lap-publishing.com

Herstellung: siehe letzte Seite /
Printed at: see last page
ISBN: 978-3-659-61211-4

# Contents

# Chapter 1.
# Numerical Model of Hardening Concrete

## 1.1. Introduction

Cement and water blending generates chemical reaction that is usually called hydration. Reaction of cement with water is in fact a set of chemical reactions and physical processes. After blending cement with water, reactions take place in the surface of the grains of cement and its components and some other products in the reaction dissolve in liquid phase [1.4], [1.6], [1.8], [1.12], [1.15]. Some components of cement dissolve congruently and fall into hydration. Other components dissolve incongruently – with disintegration, falling into hydrolysis [1.4], [1.6], [1.8], [1.12], [1.15]. In the analysed process, there are also reactions of synthesis between compounds created as the result of hydration or hydrolysis of separate cement components.

There is also a phenomenon of hardening of newly made products. Reaction of cement with water is a very complicated process. Mutual influence of different cement components reacting with water is often complicated by the activity of various supplements [1.4], [1.6], [1.8], [1.12], [1.15]. The analysed process is not only hydration but it is much more complicated process, often called the hardening [1.5], [1.7], [1.8] which seems to be adequate and well-founded. Despite the long-term research, it has not been possible to, unequivocally define the sort and character of chemical reactions taking place during the time of the cement materials hardening process [1.4], [1.6], [1.8], [1.12], [1.15].

## 1.2. Modelling of cement materials hardening process

Although cement materials have had more than one hundred years history, they are still a very important building material. Positive and negative experiences with their application were used in subsequent realizations. Boisterous development of material engineering, especially in the last few years, has created new possibilities also in the cement materials technology [1.6], [1.8], [1.12], [1.18]. This has contributed to the considerable increase in both: the durability of cement materials and the firmness of their structure. Those effects were possible to attain thanks to modifications of the cement materials structure that rested mainly on usage of specific chemical admixtures (super plasticizer) and mineral addition (silica fume). Modifications of the cement materials structure allow to reduce water-cement ratio and porosity. As a result one can obtain a mixture, in which there are much more active particles that are able to make structural bonds and much less capillary pores that create main structural defects.

The essence of the cement materials structure and the structural theory of concrete have been discussed in work [1.18] stressing the importance of the problems, which are connected, with the structural defects of the analysed materials. In this work the problems concerning the modelling of hardening process have been presented, with distinction of the models: technological, time-dependent, structural and thermo dynamical. Moreover, in work [1.18] the essence of the temperature function and the maturity of hardening cement materials have been defined.

4

It may appear that modelling problems have already been solved. In literature one can find many different hardening models: technological, time-dependent, structural and thermodynamical [1.6], [1.8], [1.10], [1.12], [1.18]. The notion of hardening concrete temperature function and the ripeness of concrete are also discussed in these dissertations.

The problem of mechanical properties development in concrete hardening process has a large bibliography. The results of considerations are systematized in form of functional or correlation relationships. It is possible to mark out two fundamental methods of hardening process modelling. In one of them the problems concerning the modelling of hardening process have been presented with distinction of technological and time-dependent models. In these models changes of hardening concrete strength are described by time-dependent function and parameters of binding material that had been used [1.2], [1.3], [1.7], [1.8], [1.9], [1.11], [1.12], [1.13], [1.18], [1.20] CEB-FIP Model Code MC 90 [1.21].

The second method of concrete hardening process modelling may be called the structural modelling. In this case mechanical properties of hardening concrete are described by relationships of structural parameters that depend on binding materials hydration degree [1.1], [1.5], [1.7], [1.8], [1.11], [1.12], [1.14], [1.16], [1.18].

The process of hardening of new generation concretes (with addition of super plasticizer and micro-silica) leads to form up qualitatively different structure than it take place in ordinary concretes. Mineral admixtures and chemical additives also affect the kinetics of structural processes. In present publication the structural model which enables to predict compressive strength of concrete, especially in early period of its hardening, is presented.

Seven different concrete mixtures there are presented in table 1.1, are considered.

## 1.3. General model assumption

Comprehensive specification of mechanical characteristics of hardening concrete, especially of high performance, creates necessity to adopt specific physical model.

Table 1.1. Components and physical characteristic of concrete's mixtures [1.18]

| No. | Components and physical characteristic | The type of concrete's mixture | | | | | | |
|-----|----------------------------------------|------|------|------|------|------|------|------|
| | | 1 A | 1 | 2 | 3 | 4 | 5 | 6 |
| 1. | W/(C + SF) | 0,52 | 0,52 | 0,47 | 0,42 | 0,42 | 0,37 | 0,32 |
| 2. | C [kg/m$^3$] | 340,0 | 345,0 | 363,0 | 394,0 | 320,0 | 348,0 | 388,0 |
| 3. | SF [kg/m$^3$] | – | – | – | – | 36,0 | 39,0 | 43,0 |
| 4. | SP [kg/m$^3$] | – | 4,310 | 4,540 | 4,925 | 8,900 | 9,675 | 10,781 |
| 5. | P [kg/m$^3$] | 989,0 | 982,0 | 988,0 | 985,0 | 1003,0 | 992,0 | 988,0 |
| 6. | G [kg/m$^3$] | 989,0 | 982,0 | 988,0 | 985,0 | 1003,0 | 992,0 | 988,0 |
| 7. | W [kg/m$^3$] | 177,0 | 177,0 | 168,0 | 163,0 | 144,0 | 137,0 | 132,0 |
| 8. | $\rho_B$ [kg/m$^3$] | 2495,0 | 2490,0 | 2512,0 | 2532,0 | 2515,0 | 2518,0 | 2550,0 |
| 9. | $\rho_{SB}$ [kg/m$^3$] | 2519,0 | 2514,0 | 2533,0 | 2545,0 | 2552,0 | 2564,0 | 2577,0 |
| 10. | s [-] | 0,990 | 0,990 | 0,992 | 0,995 | 0,985 | 0,982 | 0,990 |
| 11. | j [-] | 0,001 | 0,001 | 0,008 | 0,005 | 0,015 | 0,018 | 0,010 |
| 12. | $V_a$ [dm$^3$/m$^3$] | 10,0 | 10,0 | 8,0 | 5,0 | 15,0 | 18,0 | 10,0 |
| 13. | $V_e$-$B_e$ [s] | 10,5 | 7,0 | 8,0 | 8,0 | 9,5 | 10,5 | 9,0 |
| 14. | $f_{c,cube}$ [MPa] after 28 days in hydro isolated condition (18±2°C) | 50,4 | 53,5 | 63,7 | 77,80 | 77,7 | 86,4 | 93,5 |

In the table there are: W/(C+ SF)-water binder ratio, C, SF, SP (the 40% water solution of super plasticizer), P, G, W - content of cement, silica fume, super plasticizer, sand, basalt grit, water in 1 m$^3$ of concrete mixture respectively.
In the table there are as well: $\rho_B$, $\rho_{SB}$, s, j, $V_a$, $V_e$-$B_e$, fc, cube - apparent density and density of concrete mixture, tightness and cavity, volume of air pores, consistency of concrete mixture, compression strength of concretes respectively.
SP contains the remaining water in the formula W/(C + SF).

In this publication it is established, that the concrete can be treated as a composite material, where the dissipate phase – aggregate and grain of non-hydrated cement is joined by gel with dissipate pores, which makes a matrix. Assumptions which are established here can be a basis for description of destruction process, which, in the broad scope of structure's development, proceeds in the matrix area. Mechanical characteristic are given by following factors [1.4], [1.6], [1.7], [1.8], [1.11], [1.12], [1.18]: total porosity, pores size distribution, defect's existence, diversity of structure's level.

## 1.4. Parameters of concrete structure

Concrete materials, because of character of physical and chemical processes, occurred in cement grout included in them, and on the point of contact of filler's grains with a cement paste, have a porous structure.

In the hardening process of cement grout, next to capillary pores, molecular (gel) pores, directly connected with gel products, are made. Capacity of capillary pores with reference to the unit of binder's mass can be calculated from the formula [1.12]:

$$\omega_{cap} = \frac{w}{S} - \left(\omega_H + \omega_p\right) \cdot \alpha \qquad (1.1)$$

where w is initial capacity of water in the unit of cement grout's capacity [$dm^3$], $\omega_H$ is chemically tied water in the unit of binder's mass [$dm^3/kg$], $\omega_p$ is out of network water which stays in binding gel's structure with reference to the unit of binder's mass [$dm^3/kg$], $\alpha$ is degree of binder's hydration, S is binder's mass, w / s = $\omega$ is water – binder ratio.

Capacity of molecular (gel) pores with reference to the unit of binder's mass amounts [1.12]:

$$\omega_{gel} = 0,28 \cdot \alpha \cdot \left( \frac{1}{\rho_s} + \omega_H + \omega_p - V_s \right) \tag{1.2}$$

where: $\rho_s$ is binder's density, $V_s$ is a change of system's volume: water – cement with reference to the unit of cement's mass (contraction).

Structures parameters and thermo-physical characteristics of binders that compose the analysed concretes: 1A (**Plain Concrete** – **PC**), 1÷3, 4÷6 (**High Performance Concrete** – **HPC**), are presented in table 1.2. Value of degree of binder's hydration in the process of its hardening is approximated on the basis of own calorimetric research of hardening's heat by the equation [1.18], [1.19]:

$$\alpha = \exp\left[ -c \left( \ln t_a \right)^{-d} \right] \cdot \frac{Q_{max}}{Q_o} \tag{1.3}$$

where: c, d are empirically appointed parameters, $t_a$ is reduced time [h], $Q_{max}$, $Q_o$ are maximum and theoretical value of heat in binder's hardening.

Table 1.2. Parameters of structure and thermo-physical characteristics of binder [1.18].

| No | Concrete's mixture | $\omega_t = \omega_H + \omega_p$ cm³/g | $\omega_H$ cm³/g | $\omega_p$ cm³/g | $V_s$ cm³/g | $\rho_s$ g/cm³ |
|----|--------------------|------|------|------|------|------|
| 1. | 1A (**PC**) | 0,439 | 0,252 | 0,187 | 0,04690 | 3,125 |
| 2. | Concretes 1÷3 | 0,439 | 0,252 | 0,187 | 0,04690 | 3,125 |
| 3. | Concretes 4÷6 (**HPC**) | 0,395 | 0,278 | 0,117 | 0,04221 | 2,973 |

In table 1.3 equation's parameters (1.3), assigned to individual type of researched concretes are presented.

Equivalent time is described by the equation [1.18], [1.19]:

$$t_a = \int_0^t \exp\left[\frac{E_k}{R}\left(\frac{T_{(t)} - T_a}{T_{(t)}T_a}\right)\right]dt \qquad (1.4)$$

where: $E_k$ is an energy of chemical process activation, R is universal gas constant [J/molK], $T_{(t)}$ is an absolute temperature of course of reaction [K], $T_a$ is temperature of reference [K], t is time [h].

Table 1.3. Parameters of (1.3) equation [1.18]

| No | Concrete | c | d | $Q_{max}$ [kJ/kg] | $Q_o$ [kJ/kg] |
|----|----------|---|---|-------------------|---------------|
| 1. | 1A | 13.448 | 2.135 | 430 | 430 |
| 2. | 1 | 22.297 | 2.094 | 430 | 430 |
| 3. | 2 | 24.078 | 2.167 | 430 | 430 |
| 4. | 3 | 27.129 | 2.238 | 411,51 | 430 |
| 5. | 4 | 175.659 | 3.863 | 387 | 387 |
| 6. | 5 | 165.527 | 4.082 | 362,51 | 387 |
| 7. | 6 | 132.679 | 4.025 | 313,47 | 387 |

Energy of activation $E_k$ is an important parameter which characterizes influence of temperature on kinetics of structural transformation of binders process. Raised temperatures of hardening activate structural transformations process in various degrees, depending on binder's composition. Higher level of activation energy comes out in concretes which include micro silica (4÷6) in comparison with concretes without it (1A and 1÷3 concretes). For 4÷6 and 1A as well as 1÷3 concretes appropriate values of $E_k$ amount: 26 kJ/mol and 23 kJ/mol.

The measure of cement materials structure's condition during their hardening determines, put by [1.11], porosity coefficient, given by an equation:

$$x = \frac{\omega_{gel}}{\omega_{gel} + \omega_{cap} + \omega_a}$$

(1.5)

where: $\omega_{gel}$, $\omega_{cap}$, $\omega_a$ are adequately stand for gel, capillary and air pores capacity with reference to unit of binder's mass [dm$^3$/kg].

Taking into consideration equations (1.1) as well as (1.2) and providing $\omega = w / s$, we will get [1.18], [1.19]:

$$x = \frac{0{,}28 \cdot \alpha \cdot \left( \dfrac{1}{\rho_S} + \omega_H + \omega_p - V_S \right)}{0{,}28 \cdot \alpha \cdot \left( \dfrac{1}{\rho_S} + \omega_H + \omega_p - V_S \right) + \omega - \left( \omega_H + \omega_p \right) \cdot \alpha + \omega_a}$$

(1.6)

Porosity coefficient x assumes values of < 0,1 > range. For $\alpha = 1$, $\omega_a = 0$ and $\omega = \omega_H + \omega_p$, porosity coefficient x = 1, which means that hardened binder grout consists of hardened gel only.

Proprietary research on 1A, 1÷6 (comp. tab. 1.1) concrete made it also possible to identify the influence of the modification of cement matrix on its microstructure. The results of the analysis indicate that there is a close relationship between the properties of cement matrix and the porosity coefficient x. A visible influence of super plasticizer effect was observed for 1A and 1 types of concrete that are characterized by the identical values of a water-binder ratio. Super plasticizer's molecules adsorb on the surface of cement grains and lead to their deflocculation and in this way the use of cement is better [1.4], [1.5], [1.6], [1.10], [1.14], [1.15].

Participation of capillary pores in 1÷3 types of concrete decreases in a matrix structure according to decreasing value of water-binder ratio. The effect of a combined action of a super plasticizer and micro silica in concrete types 4÷6 is similar to the one which was observed in concrete types 1÷3. The effects of a cement matrix modification are especially visible in the analysis of inner microstructure. The research made by means of a scan microscope showed that microstructure of hardened cement paste in concrete 6 is very consistent, well packed and definitely less porous in comparison to the 1A concrete. The essential differences are presented in the structure of phase C – S – H of these types of concrete as well.

Fig. 1.1. A microscopic picture (magnification x 4000) of a hardened 1A cement paste (own research). There are crystal needles of phase C – S – H in porous places [1.18].

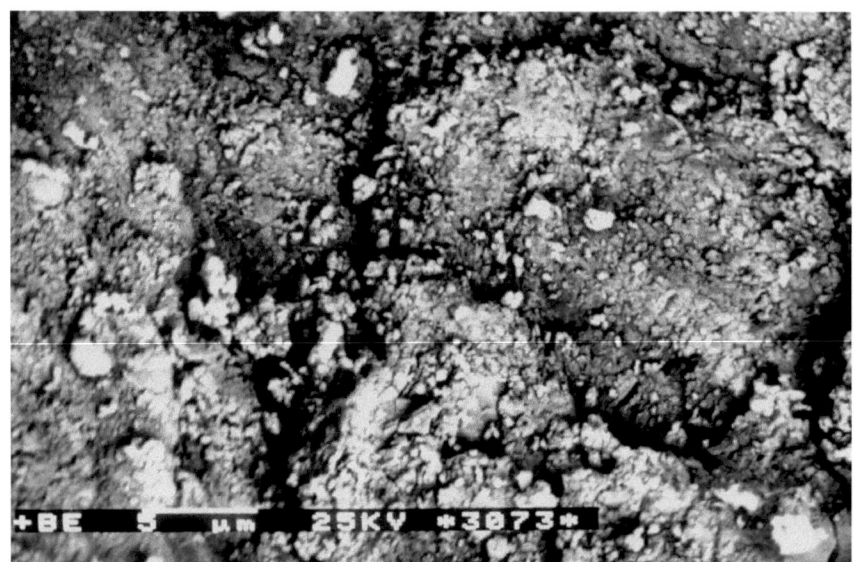

Fig. 1.2. A microscopic picture (magnification x 4000) of a hardened 6 cement paste (own research). Hydration products in the form of compact gel of the phase C – S – H [1.18].

## 1.5. Structure parameters in connection with compression strength

Mathematical model showing the dependence between concrete compressive strength and porosity coefficient of its structure (given by 1.6 formulas), assumed in this publication, is depicted by general equation [1.18], [1.19]:

$$R_c = R_o \cdot x^a \cdot \exp[b \cdot (1 - x)] \qquad (1.7)$$

where: $R_c$ is current compression strength of concretes in a given stage of structure's development, $R_o$ is theoretical compression strength of concrete's, when x = 1, a, b are empirically characterized parameters, dependent on the type of concrete's mix. The symbols $R_c$ and $R_o$ were used because strength of concrete was tested on nonstandard cubes.

General structure of formula (1.7) refers to both: conception of specification of concrete strength in porosity coefficient's function [1.11] and ceramic material's model given by [1.17]. Exponential part of formula (1.7) expresses influence of grain's size, thus pores structure, on material's strength. A and b parameters of equation (1.7) for an individual groups of concrete's blends are given by the method of multiple regression with simultaneous definition of correlation coefficient. The results of computations for individual groups of concrete are depicted on fig. 1.3, fig. 1.4 and fig. 1.5.

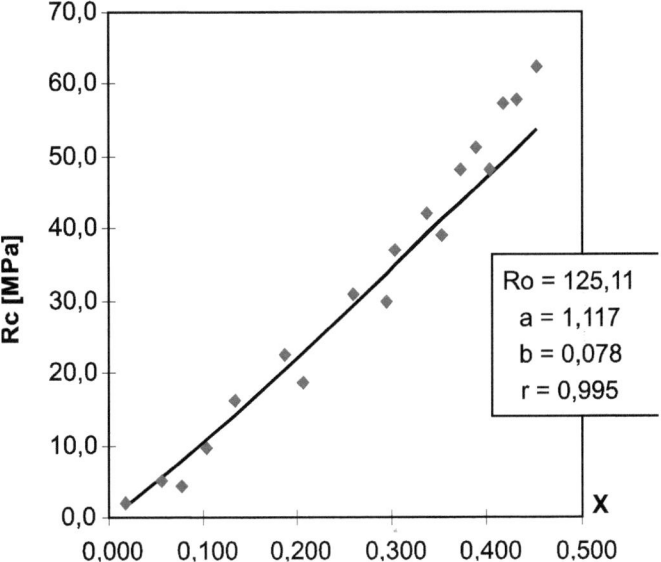

Fig. 1.3. Graph of formula (1.7) for concrete 1A [1.18], [1.19].

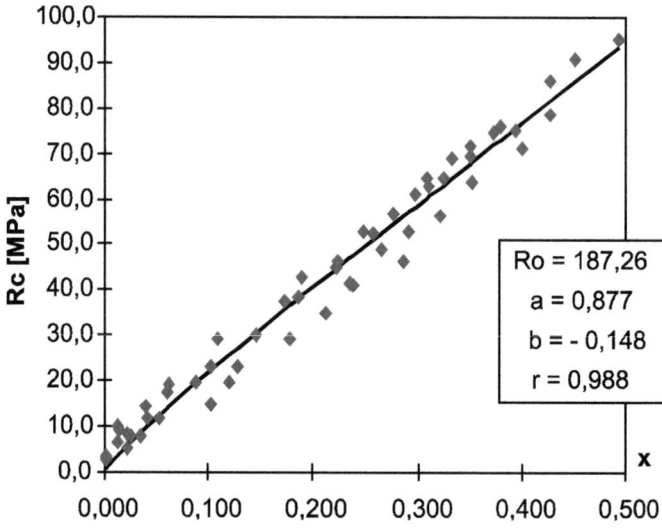

Fig. 1.4. Graph of formula (1.7) for concretes 1, 2, 3 [1.18].

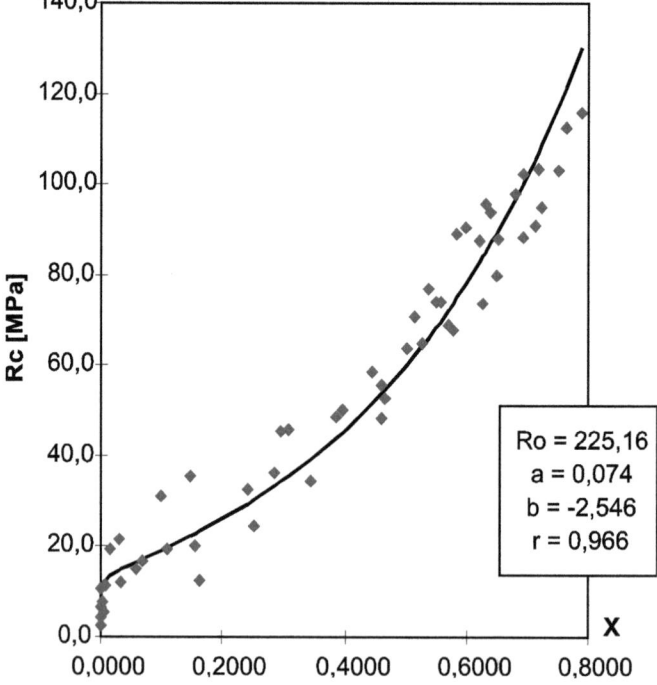

Fig. 1.5. Graph of formula (1.7) for concretes 4, 5, 6 [1.18], [1.19].

14

## 1.6. Conclusions

Formulas, shown on the basis of established model assumptions, permit expectations of development of concrete's strength under compression in wide range of changes of its structure, defined by porosity coefficient. This coefficient, giving the ratios between molecular (gel) pores capacity and total pores capacity in hardening concrete, is good description of structures – formed processes character, integrally connected with structure of porosity of hardening binding gel. The analyse of graphs shown on figures nr 1.3, 1.4 and 1.5 permit statement that various concretes show various characteristics at the same value of porosity coefficient. It means that there is an influence of mineral additives and physical and chemical active admixtures on development of strength of hardening concrete. In the established model the fundamental influence on concrete's strength in the period of its structural changes, exerts the cement gel with dissipated, molecular, capillary and air pores.

## References to chapter 1

[1.1] Bergström S.G.: Byggnadsmateriallära, (Building Materials), FKI. Lunds Tekniska Högskola, VBV, Lund, 1967.

[1.2] Byfors J.: Plain concrete of early ages. Swedish Cement and Concrete Research Institute, Of 3, Stockholm.

[1.3] Bresson J.: Prediction of strength of concrete products. RILEM International Conference on Concrete at Early Ages, Paris, 1982, vol. I, p.111-115.

[1.4] Czarnecki I., Broniewski T., Henning O.: Chemia w budownictwie. (Chemistry in Civil Engineering). Arkady, Warszawa 1994.

[1.5] Hansen T.C.: Physical Structure of hardened Cement Paste-A Classical Approach. Materials and Structure, 19, 114, 1986, p. 423 - 436.

[1.6] Jasiczak J., Mikołajczyk P.: Technologia betonu modyfikowanego domieszkami I dodatkami. Przegląd tendencji krajowych I zagranicznych. (Technology of concrete modified by additives and admixtures. Overview of domestic and foreign trends). Wydawnictwo Politechniki Poznańskiej, Poznań 1997.

[1.7] Kiernożycki W.: Compressive and shearing strength of hardening concrete. Archives of Civil Engineering, XXXIX, 1, 1993, p. 61-76.

[1.8] Kiernożycki W.: Betonowe konstrukcje masywne. (Massive Concrete Structures). Polski Cement, Kraków 2003.

[1.9] Kishi T., Shimomura T., Maekawa K.: Thermal crack control design of HPC. Proceedings of the International Symposium held by RILEM, E&FN SPON, Belgium 1991.

[1.10] Kucharska L.: Kształtowanie struktury wysokosprawnych betonów. Rola dodatków I domieszek. (Shaping the structure of high-performance concrete. The role of additives and admixtures). Przegląd Budowlany nr 8-9/1992.

[1.11] Mikoś J.: Wytrzymałość betonu w funkcji współczynnika porowatości. (Concrete strenght as a function of porosity ratio). Archiwum Inżynierii Lądowej. Tom XXX z. 1-2/1985.

[1.12] Neville A.M.: Properties of Concrete. Pearson Education Limitem. Fourth Edition, 1995.

[1.13] Popowics S.: Generalizations of the Abram's law-Prediction of strength developments of concrete from cement properties. ACI Journal, 78, 1981, p.123-129.

[1.14] Powers T.C.: The physical structure and engineering properties of concrete. Portl. Cem. Assoc. Res. Dept. Bull. 90, 1958.

[1.15] Reul H.: Handbuch der Bauchemie. (Handbook of Chemistry in Civil Engineering). Verlag für Chemische Industrie. Ziolkowsky AG, Augsburg 1991.

[1.16] Rostasy F.S, Laube M., Onken P.: Zur kontrolle früher Temperaturrisse in Betonbauteilen. (Kontrola wczesnych zarysowań termicznych w elementach betonowych). Bauingenieur, 68, 1993, s. 5-14.

[1.17] Spriggs R.M., Vasilos T.: Effect of grain size on transverse bond strength of alumine and magnesia. J.AM.Ceram.Soc.46,224,1963.

[1.18] Ślusarek J.: Model twardnienia tworzyw cementowych. (The cement materials hardening model). Zeszyty Naukowe Politechniki Śląskiej Nr 1513, seria Budownictwo, z. 92, Wydawnictwo Politechniki Śląskiej, Gliwice 2001, stron 141.

[1.19] Ślusarek J.: Hardening concrete effects in massive structures. EJPAU-CIVIL ENGINEERING (Electronic Journal of Polish Agricultural Universities), 2008, 11(2), #06.

[1.20] Weber J.W.: Empirische Formeln zur beschreibung der Festigkeitentwicklung und der Entwicklung des E-Modulus von Beton. (Empiryczne formuły do opisu rozwoju wytrzymałości I modułu sprężystości betonu). Betonwerk+Fertigteil-Technik, 12, 1979, p. 753-756.

[1.21] CEB-Comite' Euro-International du Beton-CEB-FIP Model Code 1990. Bulletin D' Information, 213/214, Lausanne, 1993.

# Chapter 2.
## Temperature Fields In Experimental Blocks

## 2.1. Introduction

The setting and hardening of concrete is accompanied by nonlinear temperature distributions caused by the heat of hydration [2.1]÷[2.4], [2.6], [2.7], [2.14], [2.16], [2.17]. High temperature gradients associated with the exothermic chemical reactions of cement hydration may occur between the interior and the surface of structural elements at early ages [2.3], [2.7], [2.14]÷[2.16]. Cracks occur when these temperature gradients lead to tensile stresses that exceed the tensile strength of the young concrete, influencing the durability of the structure [2.7], [2.10], [2.11], [2.16]. This problem is especially accentuated in massive concrete structures [2.1]÷[2.4], [2.6], [2.7], [2.10], [2.11], [2.15], [2.16], [2.17]. In the case of High-Performance-Concrete structures, such effects can also be significant, due to the higher cement and silica-fume contents [2.3], [2.4], [2.7].

In this publication, a theoretical and experimental study about temperature distributions induced in plain and high-performance concrete structures during the hardening process are presented. Heat of hydration effects were studied for the two large concrete blocks. One of them was made of ordinary concrete with water to binder ratios 0.52 (**PC**). The second block was made of high – performance concrete with water to binder ratios 0.32 (**HPC**). The concrete blocks had a cylindrical form 100 cm in diameter. They were waterproof and thermally insulated from above and below.

## 2.2. General model assumption

Assuming that concrete verifies several hypotheses – continuum, isotropy and homogeneity – the well-known differential equation that governs the heat transfer problem in these cases is

$$\frac{\partial^2 T}{\partial r^2} + \frac{1}{r} \cdot \frac{\partial T}{\partial r} = \frac{c \cdot \rho}{\lambda} \cdot \frac{\partial T}{\partial r} - \frac{q_v}{\lambda}, \qquad (2.1)$$

in which $\lambda$ is the thermal conductivity [W/m·K]; c is the specific heat [J/kg·K]; $\rho$ is the density [kg/m$^3$]; $q_v$ is the rate of heat generated per volume unit [W/m$^3$]; r is the radius of the cylinders [m]; T is the temperature [K] and t means time [s]. Replacing the partial derivatives with the differential quotients and making some transformations we can obtain an well-known equation

$$T_{i,k+1} = T_{i,k} \frac{\lambda}{c\rho} \cdot \frac{\Delta t}{\Delta r^2} \left[ \left(2 + \frac{1}{i}\right) T_{i,k} - \left(1 + \frac{1}{i}\right) T_{i+1,k} - T_{i-1,k} \right] + c\rho \Delta t \cdot qv_{i,k}, \quad (2.2)$$

Basing on an equation (2.2) we can calculate the temperature fields in time $(k+1)\Delta t$ assuming that the temperature fields in time $k\Delta t$ are known. The energy transferred between the boundary and the environment may be expressed by Newton's law

$$\alpha_p \left[ T_n - T_f \right] = -\lambda \frac{\partial T}{\partial r} \bigg| n, \qquad (2.3)$$

in which $\alpha_p$ is the boundary heat transfer coefficient [W/m$^2$K]; $T_n$ is the boundary temperature [K]; $T_f$ is the environment temperature [K].

Replacing the partial derivative $\dfrac{\partial T}{\partial r}$ with the differential quotient and making some transformations we may express the boundary temperature of the analyzed cylinders by the following formula

$$T_n = \frac{\alpha_p \cdot \Delta r \cdot T_f + \lambda \cdot T_{n-1}}{\alpha_p \cdot \Delta r + \lambda}. \tag{2.4}$$

## 2.3. Internal heat sources function

To obtain the nonlinear temperature fields in large concrete columns it was necessary to determine the internal heat sources function $W(t)$ connected with the heat of concrete hardening $Q_t$ which can be expressed by the equation

$$Q_t = \int_0^t W(t)dt. \tag{2.5}$$

In literature there are many formulas to making possible identifying the internal heat sources of hardening concrete. There are estimated on the bases of results of isothermal or adiabatic hardening process [2.7], [2.8], [2.13]. For the isothermal conditions there is possible to find the exponential function [2.9], [2.15], the form of parabola fragment [2.15] and in the form of polynomial [2.8]. For the adiabatic conditions of hardening concrete the internal heat sources may be approximated by exponential function [2.6] as well. In work [2.16] we can find the heat sources equation in the form of difference of two exponential functions. Very used in practice interesting algorithm was carried out in work [2.17]. The internal heat sources function may be appointed on the bases of known isothermal functions series. There is possible to obtain this

20

function on the bases of isothermal researches [2.17]. Influence of temperature on power of internal heat sources was carried out in work [2.5]. There is possible to define of this influence on the bases of temperature function, which may be given for example in the form of exponential function [2.5].

In this article heat of hydration effects in **PC** and **HPC** mixes (table 1 – chapter 1 of this publication) were studied by means of BMR differential micro calorimeter. Based on experimental research, realized in temperatures 283 K, 298 K, 313 K, the internal heat sources function W(t) can be determined by the following formula [2.1]

$$W(t) = Q_{max} \frac{p \cdot q \cdot exp\left[-p(\ln t\alpha)^{-q}\right]}{t(\ln t\alpha)^{q+1}}, \tag{2.6}$$

where: $Q_{max}$ is the maximum value of binder hydration heat [kJ/kg]; p, q are the parameters described in table 2.1; the variable ta is an equivalent time of the process [2.7], [2.16] obtained by

$$T_a = \int_0^t exp\left[\frac{E_k}{R}\left(\frac{T_{(t)} - T_a}{T_{(t)} \cdot T_a}\right)\right] dt, \tag{2.7}$$

where: $E_k$ is an activation energy of the chemical process presented in table 2.2 [kJ/mol]; Ta is a reference uniform temperature [K] and R is an universal gas constant [kJ/mol·K]; t means time [h].

The function described by formula (2.6) is monotonically decrease and at the beginning of concrete hardening process is not good approximating of reality. Independently of this fact on the bases of experience of Authors of the work [2.1] this formula was accepted arbitrarily for the next investigations.

Table 2.1. Thermo physical properties of concretes

| Type of concrete[1] | Parameters | | | | | | | |
|---|---|---|---|---|---|---|---|---|
| | $c^{2)}$ | p | q | $Q_{max}$ | $Q_o$ | $\lambda^{2)}$ | $E_k$ | $\alpha_p$ |
| PC | 0.971 | 13.448 | 2.135 | 430 | 430 | 2.195 | 23 | 5 |
| HSC | 0.908 | 132.679 | 4.025 | 313.47 | 387 | 2.184 | 26 | 5 |

1) Composition and physical properties of the mixes are given in table 2.1
2) Parameters calculated on basis of Kiernożycki W. [2.8]

## 2.4. Temperature distributions in the experimental blocks

Concrete hardening temperatures of the tested columns were estimated on basis of equations (2.2) and (2.4)

$$q_{vi} = W_i(t) \cdot B,\tag{2.8}$$

where: B is binders content [kg/m$^3$].

Examinations of concrete hardening temperatures were realized in two experimental blocks. In the researches supervisory measuring equipments with continually recording system of temperature fields were used. Locations of measuring points in analysed experimental blocks are shown on fig. 2.1. Registered temperature distributions in the **PC** and **HPC** blocks are presented in fig. 2.2 and fig. 2.3. Theoretical and experimental surface temperature distributions in the analysed blocks are presented in fig. 2.4 and fig. 2.5.

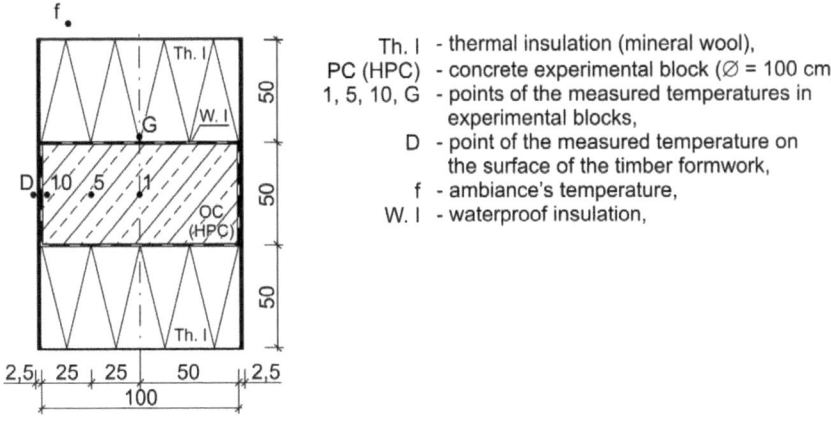

Th. I - thermal insulation (mineral wool),
PC (HPC) - concrete experimental block (∅ = 100 cm),
1, 5, 10, G - points of the measured temperatures in
experimental blocks,
D - point of the measured temperature on
the surface of the timber formwork,
f - ambiance's temperature,
W. I - waterproof insulation,

Fig. 2.1. Location of measuring points in analyzed experimental blocks

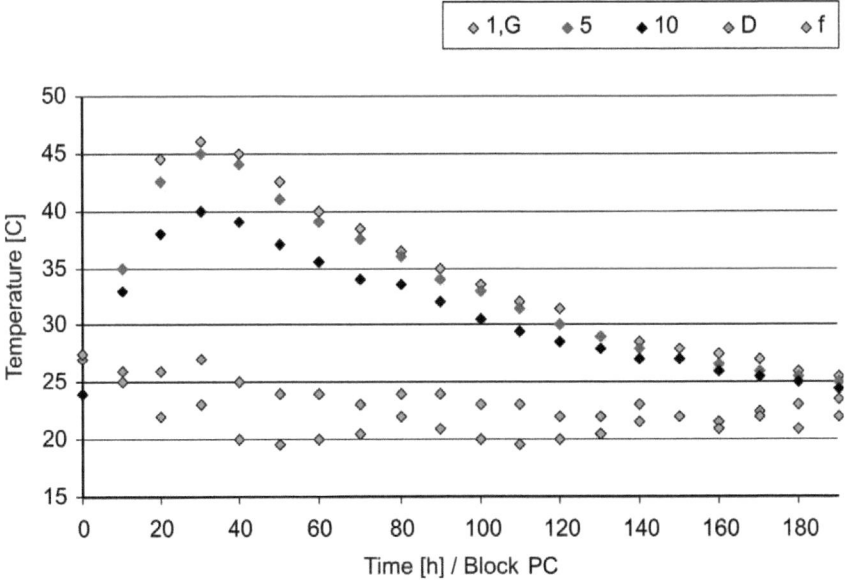

Fig. 2.2. Registered temperature distributions in the **PC** block

23

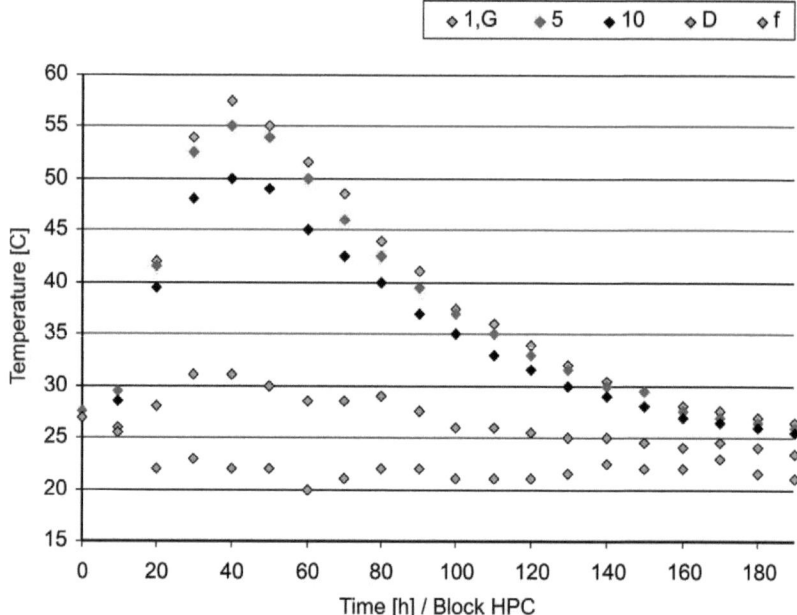

Fig. 2.3. Registered temperature distributions in the **HPC** block

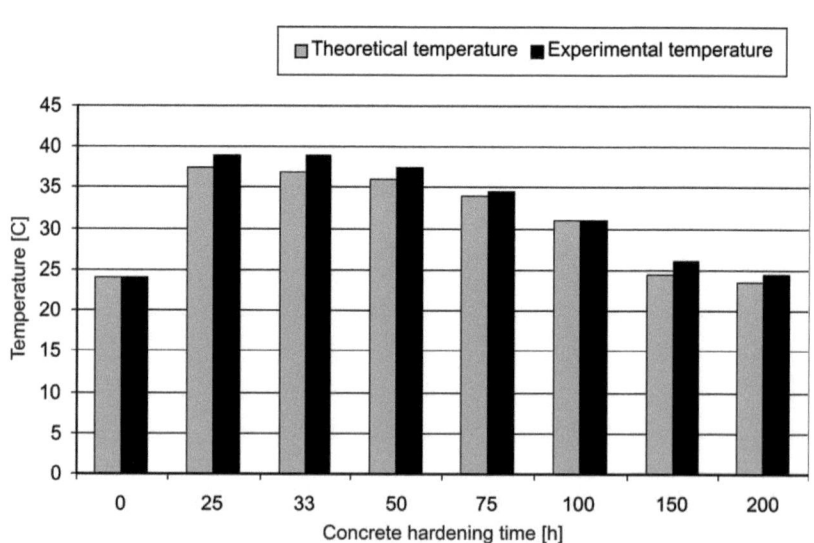

Fig. 2.4. Surface temperature of hardening **PC** block, $T_{max}$ after 25 h

24

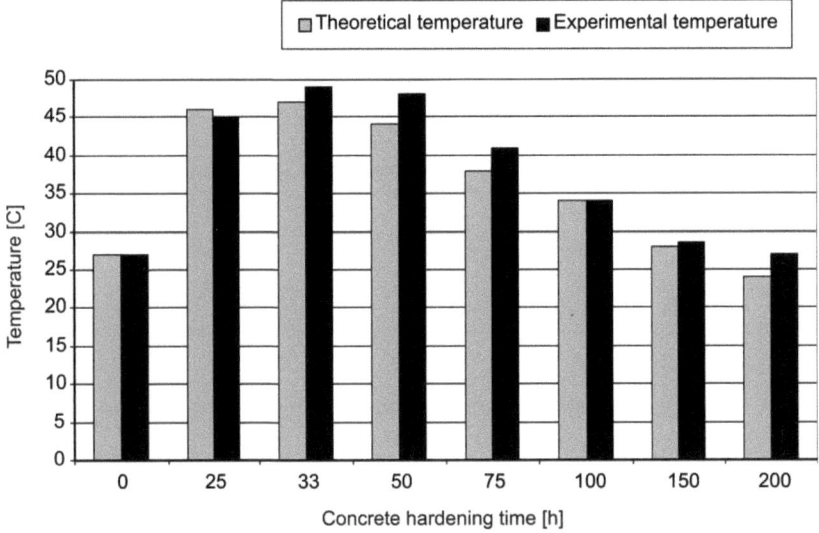

Fig. 2.5. Surface temperature of hardening **HPC** block, $T_{max}$ after 33 h

## 2.5. Conclusions

The setting and hardening of concrete is accompanied by nonlinear temperature distributions caused by the hydration heat development. Heat of hydration effects in large **PC** and **HPC** blocks where tested by experimental way. On the bases of Finite Difference Method the simulation model for development of temperature distributions of the investigated large concrete blocks was carried out. Results of the theoretical way were tested as well. The results obtained by experimental and theoretical way are comparable.

## References to chapter 2

[2.1]  Branco F. A., Mendes P. A., Mirambel E.: Heat of hydration effects in concrete structures. ACI Materials Journal, v. 89, 2/1992.

[2.2] Eierle B., Schikora K.: Badensplatten unter frühem Temperaturzwang – Rechenmodelle und Tragverhalten. Bauingenieur, 75, 200, p. 671-678.

[2.3] Ekerfors K., Jonasson I. E., Emborg M.: Behaviour of Young HSC. Utilization of HSC. 20÷24 June1993, Lillehamer, Norway, v. 20.

[2.4] Emborg M.: Thermal Stresses in Concrete Structures of Early Ages. Lulea University of Technology,1989: 73 D.

[2.5] Flaga K.: Influence of concrete self heating on erection speed of massive structures. Inżynieria i Budownictwo 3/1970, p. 96-98 [in Polish].

[2.6] Hirschfeld K.: Die Temperaturvertailung im Beton. Springer Verlag, Berlin Göttingen-Heidelberg 1948

[2.7] Kiernożycki W.: Massive concrete structures. Polski Cement, Kraków 2003 [in Polish].

[2.8] Kurzawa J., Kiernożycki W.: Uwarunkowania technologiczne w procesie realizacji elementów I konstrukcji masywnych z betonu. Prace Naukowe Politechniki Szczecińskiej nr 441, Szczecin, 1991 [in Polish].

[2.9] Rastrup E.: Heat of hydration in concrete. Magazine of Concrete Institute, 1947.

[2.10] Rostasy F. S., Henning W.: Zwang in Stahlbetonwän auf Fundamenten. Beton- und Stahlbetonbau, 8, 9, 1989, p. 208-214, 232-237.

[2.11] Simons H. J.: Betonierabschnitte von Stahlbetonboden-platen ohne Mindertbewehrung. Beton-und Stahlbetonbau, 94, 6, 1999, p. 254-258.

[2.12] Setzer J.: A model of hardened cement paste for linking shrinkage and creep phenomena. Matinus Nijhoff Publishers, Haque/Boston/London, 1982.

[2.13] Ślusarek J.: Heat of hydration effects in large high – strength concrete columns. Proceedings of 10th International Symposium for Building Physics, Dresden 27 ÷ 29, September 1999, p. 743-752.

[2.14] Van Breugel K.: Simulation model for development properties of early age concrete. Symposium RILEM, Ghent 07.1999.

[2.15] Weshe K.: Baustoffe für tragende Bauteile. Wiesbaden, Bauverlag, 1993

[2.16] Witakowski P.: Analysis of thermal stresses in large concrete blocks. WKiŁ, Warszawa, 1977 [in Polish].

[2.17] Witakowski P.: Thermodynamic theory of maturing. Application to massive concrete structures. Politechnika Krakowska, IL z.70, Zeszyt Naukowy 1, Kraków 1998 [in Polish].

# Chapter 3.
# Exertion of experimental concrete blocks
# caused by heat of hydration

## 3.1. Introduction

The setting and hardening of concrete is accompanied by nonlinear temperature distributions caused by the heat of hydration [3.1]÷[3.4], [3.6], [3.7], [3.14], [3.16]. High temperature gradients associated with the exothermic chemical reactions of cement hydration may occur between the interior and the surface of structural elements at early ages [3.3], [3.7], [3.14]÷[3.16]. Cracks occur when these temperature gradients lead to tensile stresses that exceed the tensile strength of the young concrete, influencing the durability of the structure [3.7], [3.10], [3.11], [3.16]. This problem is especially accentuated in massive concrete structures [3.1]÷[3.4], [3.6], [3.7], [3.10], [3.11], [3.15], [3.16]. In the case of High-Performance-Concrete structures, such effects can also be significant, due to the higher cement and silica-fume contents [3.3], [3.4], [3.7].

In this publication, a theoretical and experimental study about temperature and stress distributions induced in ordinary and high-performance concrete structures during the hardening process are presented. Heat of hydration effects were studied for the two large concrete blocks. One of them was made of ordinary concrete with water to binder ratios 0.52. The second block was made of high – performance concrete with water to binder rations 0.32. Composition and properties of the used in the experimental tests are presented in table 3.1. The concrete blocks had a cylindrical form 100 cm in diameter. They were waterproof and thermally insulated from above and below.

## 3.2. Thermal stresses in the experimental cylindrical blocks – basic equations

In the publication examples of numerical analysis of concrete blocks thermal stresses are presented. The stresses of external surface of endless cylinder were observed. Thermal stresses in the cylindrical experimental blocks can be expressed by equations [3.13]:

$$\sigma_r = \frac{\alpha_T E}{1-v}\left(\frac{1}{b^2}\int\limits_0^b Trdr - \frac{1}{r^2}\int\limits_0^r Trdr\right), \tag{3.1}$$

$$\sigma_\theta = \frac{\alpha_T E}{1-v}\left(\frac{1}{b^2}\int\limits_0^b Trdr + \frac{1}{r^2}\int\limits_0^r Trdr - T\right), \tag{3.2}$$

$$\sigma_z = \frac{\alpha_T E}{1-v}\left(\frac{2}{b^2}\int\limits_0^b Trdr - T\right). \tag{3.3}$$

In this case we are analysing thermal stresses on the external surface of the cylinders, therefore $\sigma_r = 0$ and $\sigma_\theta = \sigma_z$.

In these analyses there is no possible to observed an inversion of stresses effects. This effect, of course, must not takes a place in the point of cylinder external surface.

Analyses of the tested blocks exertion can be made on the grounds of Zandel's hypothesis described in [3.5]

$$\sigma_1 + \left(1 - \frac{R_r}{R_c}\right)\frac{\sigma_2}{2} - \frac{R_r}{R_c}\sigma_3 = R_r, \qquad (3.4)$$

where: $\sigma_1 \geq \sigma_2 \geq \sigma_3$ – the main stresses [MPa], $R_r$, $R_c$ – the tensile and compressive strength of concrete, respectively [MPa].

Based on formula (3.4), the exertion function can be expressed as follows

$$W = \frac{\sigma_1}{R_r} + \left(1 + \frac{R_r}{R_c}\right)\frac{\sigma_2}{2 \cdot R_r} - \frac{1}{R_c}\sigma_3, \qquad (3.5)$$

For the external cylinder surface, when $\sigma_1 = \sigma_2 = \sigma$ and $\sigma_3 = 0$, the exertion function can be described by the equation:

$$W = \frac{\sigma}{R_r}\left(1.5 - 0.5\frac{R_r}{R_c}\right). \qquad (3.6)$$

Analysis of thermal stresses caused by exothermic of set-up process requires to take into account rheology phenomenon in hardening concrete inside massive structure. Euro code 2 [3.17] estimating after effects of rheology distortions, recommends usage of equation of concrete state (shape) in algebraic form:

$$\varepsilon_{tot}(t,t_0) = \varepsilon_n(t) + \sigma(t_0) \cdot J(t,t_0) + \Delta\sigma(t,t_0) \cdot \left[\frac{1}{E_c(t_0)} + k\frac{\Phi(t,t_0)}{E_{c28}}\right], \qquad (3.7)$$

where: $\varepsilon_{tot}(t,t_0)$ – total deformation of concrete; $\varepsilon_n(t)$ – independent on stresses, forced deformation; $t_0$ – time of initial load of concrete; $t$ – time of calculations; $J(t,t_0)$ – the creep function after time t; $E_c(t_0)$ – the modulus of elasticity for concrete after time to; $E_{c28}$ – the modulus of elasticity for concrete after 28 days; $\Phi(t,t_0)$ – the creep coefficient; $k$ – the ageing coefficient; $\sigma(t_0)$ – stress at time to; $\Delta\sigma(t,t_0)$ – change of stress at $(t-t_0)$.

Equation (3.7) in case where solid value of deformation of producing concrete at to is kept, stress $\sigma(t_0)$, enables fixing of stress changing in relaxation process:

$$\Delta\sigma = \sigma(t_0) - \sigma(t) = \sigma(t_0) \cdot \frac{\Phi(t,t_0)\dfrac{E_c(t_0)}{E_{c28}}}{1 + k \cdot \Phi(t,t_0)\dfrac{E_c(t_0)}{E_{c28}}}. \tag{3.8}$$

Relational change of stress with time is defined by relaxation coefficient:

$$\Psi(t,t_0) = \frac{\Phi(t,t_0)\dfrac{E_c(t_0)}{E_{c28}}}{1 + k \cdot \Phi(t,t_0)\dfrac{E_c(t_0)}{E_{c28}}}. \tag{3.9}$$

Thermal stress, with regard to rheological deformation can be expressed by the equation

$$\sigma(t) = \sigma(t_0) \cdot [1 - \Psi(t,t_0)], \tag{3.10}$$

where: $\Psi(t,t_0)$ is the relaxation factor.

## 3.3. Algorithm of calculations of temperature fields and exertion of hardening concrete

Application of superposition rule of influence, occurring in the various age of concrete in consecution of temperature changing in particular steps of time at use of equation (3.9) enables to estimate total thermal stresses in the point given at massive structure. Temperature distribution of hardening concrete in experimental blocks obtained earlier as well as structural and thermo physical parameters of applied building compounds defined before at usage of derived mathematical dependence let to estimate the exertion of hardening concrete according to algorithm given in fig. 3.1.

In these publication examples of numerical analyses of concrete blocks exertion are presented. The exertion of external surface of endless cylinder $W_{10}(t)$ was observed. In presented algorithm material performances were taken into consideration on the bases of proper and literature investigations. The relationship between strength of concrete and modulus of concrete elasticity can be calculated on the basis of work [3.9]:

$$E = 4274 \cdot \rho_b \cdot (R_c)^{\frac{1}{3}}, \tag{3.11}$$

where $\rho_b$ is the apparent density of concrete.

Using proper researches on the basis of equation (3.11) we obtain:

for **PC**:

$$E = 10664 \cdot (R_c)^{\frac{1}{3}}, \tag{3.12}$$

for **HPC**:

$$E = 10889 \cdot (R_c)^{\frac{1}{3}}.$$ (3.13)

The dependence of concrete tension strength on its compression strength can be calculated, using proper researches, on the basis of the following equation:

for **PC**:

$$E = 0,070 \cdot (R_c)^{1,095},$$ (3.14)

for **HPC**:

$$E = 0,197 \cdot (R_c)^{0,748}.$$ (3.15)

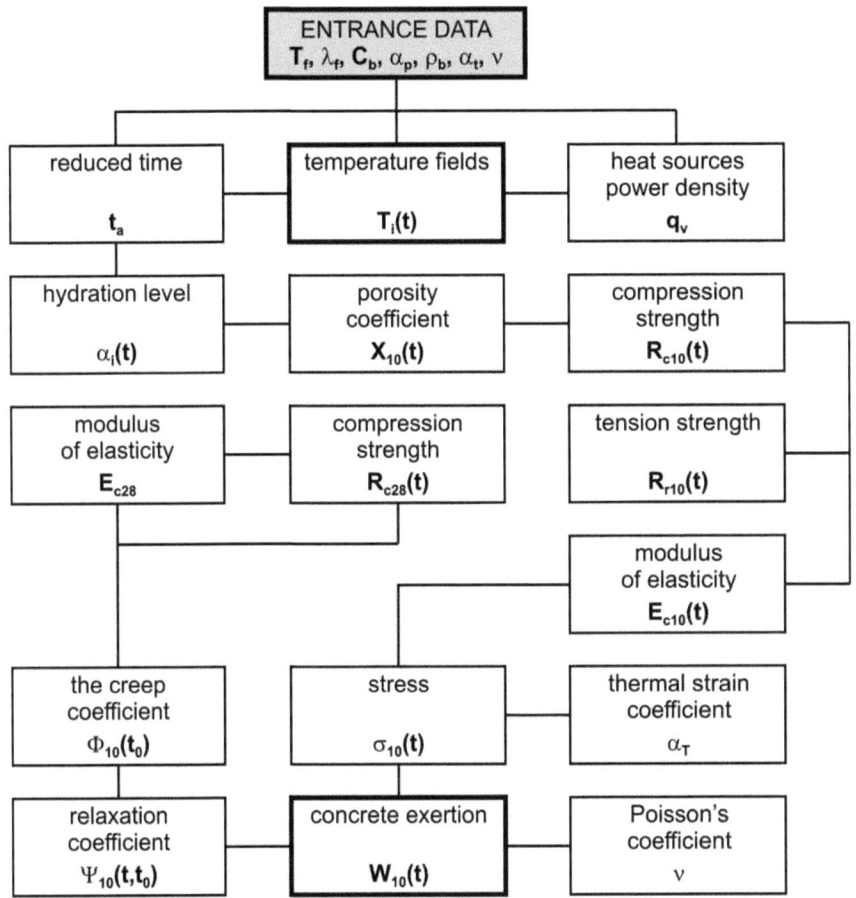

Fig. 3.1. Algorithm of calculations of temperature field and exertion of hardening concrete

## 3.4. Analysis of test results of experimental blocks exertion

Based on equations and parameters described earlier, the distribution of exertion function for **PC** and **HPC** hardening process can be calculated. Results given by computer program are presented in figures 3.2 and 3.3.

Numerical analysis of exertion of hardening concrete in experimental blocks (**PC**, **HPC**) showed dissimilar process of exertion function.

Maximum value of concrete exertion in blocks **PC** occurred after passage of 17 hours from the moment of its mould. In (particular) case of hardening concrete in block **HPC** analogous time amount to 32 hours. Level of maximal exertion of hardening concrete in analyzed experimental blocks can be recognize as comparable. At this moment it has to be marked out that initial temperature of concrete compound placed in those blocks was different and for **PC** concrete was $T_p = 24°C$, however for **HPC** concrete $T_p = 27°C$.

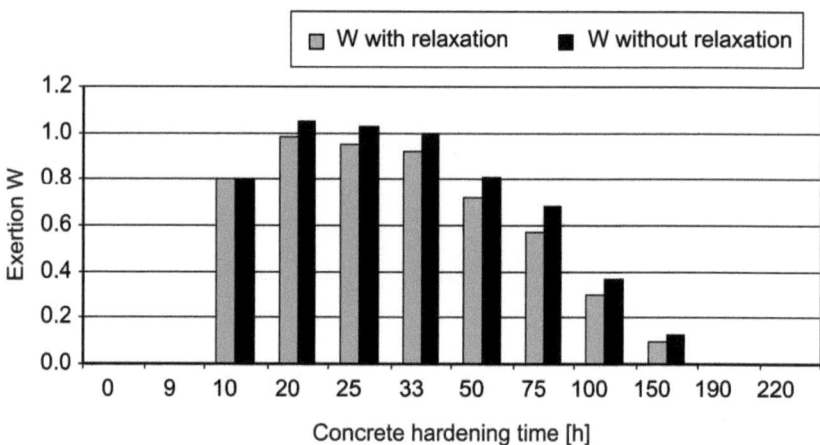

Fig. 3.2. Graph of the exertion on the external cylinder surface – function for **PC**

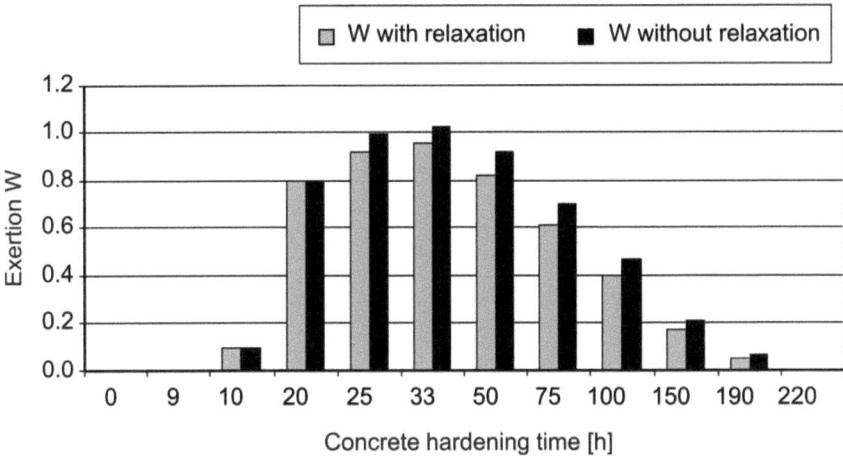

Fig. 3.3. Graph of the exertion on the external cylinder surface – function for **HPC**

In table 3.1 results of numerical analysis made for tested concrete with usage of cylinder model where diameter equals 1.0 m were showed. $T_{max}$ is maximal temperature reached by hardening concrete as a result of exothermic of structure-construction process with time $t_{\Delta T}$, after which maximal self-warm-up of concrete $\Delta T$. $W_{max}$ indicates maximal power of exertion reached by concrete in $t_{Wmax}$ time. Calculation were made with assumption that external ambient temperature $T_f$ is even during whole process of hardening and equals initial temperature of concrete blend $T_p$ = 20°C. Coefficient of taking over heat from outer cylinder surface is assumed as $\alpha_p$ = 5 W/m²K.

Table 3.1. Results of numerical calculations for various concretes

| Initial temperature of concrete blend $T_p$ [°C] | Type of concrete | Water binder indicator W/S [-] | Self-warm-up of concrete (inside solid) | | | Exertion (solid surface) | |
|---|---|---|---|---|---|---|---|
| | | | $T_{max}$ [°C] | $\Delta T$ [°C] | $t_{\Delta T}$ [h] | $W_{max}$ [-] | $t_{w,max}$ [h] |
| $T_p = T_f =$ 20°C | **PC** | 0.52 | 44.8 | 24.8 | 65.5 | 0.51 | 25.0 |
| | **HPC** | 0.32 | 50.3 | 30.3 | 62.8 | 0.54 | 42.3 |

On the basis of table 3.1 it can be said that exertion of ordinary concrete (**PC**) and high-performance-concrete (**HPC**) is comparable. In each of tested concrete it is stated that maximal value of exertion (on the solid surface) appears sooner than maximal value of self-warm-up inside solid. Maximal exertion of concrete is only slight size depends on relation between water and binder (W/S). Results of calculations made on this field for different concretes were showed in table 3.2.

Table 3.2. Dependence of exertion of ordinary concretes on water-cement ratio

| Water-cement relation | Temperature | | Exertion (solid surface) $W_{max}$ [-] |
|---|---|---|---|
| | initial of concrete mix $T_p$ [°C] | ambient $T_f$ [°C] | |
| 0.52 | | | 0.51 |
| 0.57 | 20 | 20 | 0.51 |
| 0.62 | | | 0.50 |
| 0.67 | | | 0.49 |

Profitable effects, revealed by lowering maximal level of concrete exertion hardening inside solid construction, can be reached by initial lowering of initial temperature of concrete blend to ambient temperature. Results of these calculations made for **PC** and **HPC** are showed in table 3.3.

Table 3.3. Dependence of exertion **PC** and **HPC** on cooling (down) concrete mix

| Initial temperature of concrete compound $T_p$ [°C] | Ambient temperature $T_f$ [°C] | Exertion solid surface $W_{max}$ [-] | |
|---|---|---|---|
| | | PC | HPC |
| | 10 | 0.66 | 0.67 |
| | 15 | 0.63 | 0.66 |
| 10 | 20 | 0.60 | 0.65 |
| | 25 | 0.58 | 0.64 |
| | 30 | 0.56 | 0.63 |
| | 35 | 0.55 | 0.62 |

In figures 3.11 and 3.12 cylinder (diameter 1.0 m) exertion function charts were showed. Cylinders were made of **PC** and **HPC** hardening at ambient temperature of $T_f$ = 20°C. Initial temperature of concrete compounds both **PC** and **HPC** was equal to ambient temperature. On the basis of analysis of fig. 3.10 it can be said that **PC** and **BWW** reach comparable value of maximal exertion, however in different time. Lag (delay) regarding to **PC** observed in **HPC** to gain maximal exertion, is caused by lagging effect (influence) of super plasticizer, present in **HPC** composition. On the basis of analysis of calculations result, slight influence of rheological distortion of hardening concrete in massive construction was stated on the level of its maximal exertion. Cutback (reduction) of maximal exertion on account of action of rheological distortion is similar in case **PC** and **HPC** and it's about 10%. Influence of rheological distortion on exertion of hardening concrete is tested massive construction is disclosed more intensively with time. Results of calculations brought through in this range for **PC** and **HPC** are showed in table 3.4.

Table 3.4. Influence of observation time in rheological distortion action intensity in concrete exertion

| W = exertion with relaxation / W = exertion without relaxation | | | | | | | | |
|---|---|---|---|---|---|---|---|---|
| For **PC** after time [hour] | | | | for **HPC** after time [hour] | | | | $T_p = T_f$ |
| 100 | 150 | 200 | 250 | 100 | 150 | 200 | 250 | |
| 0.83 | 0.77 | 0.73 | 0.65 | 0.86 | 0.79 | 0.75 | 0.67 | 20°C |

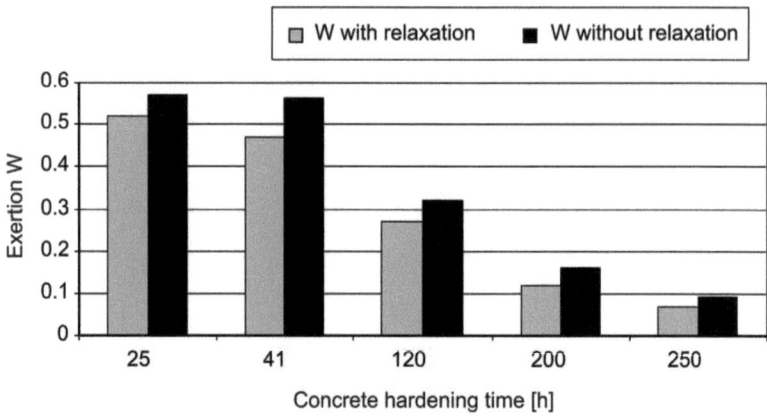

Fig. 3.4. Graph of the exertion function for **PC** block ($T_p = T_f = 20°C$), $W_{max}$ after 25 hours

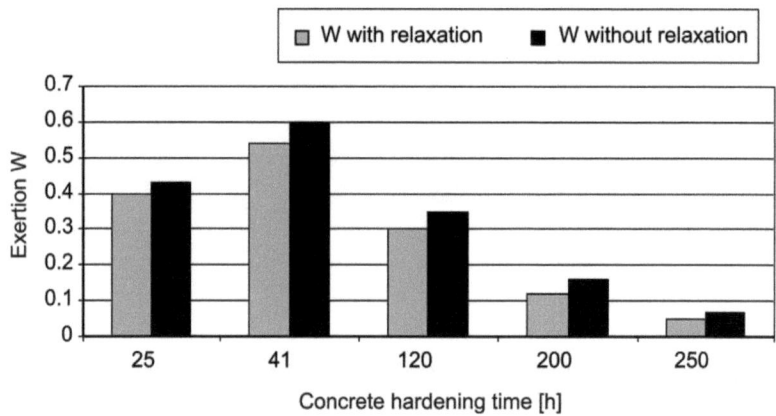

Fig. 3.5. Graph of the exertion function for **HPC** block ($T_p = T_f = 20°C$), $W_{max}$ after 41 hours

## 3.5. Conclusions

A simulation model of early age concrete is presented for prediction of development of the degree of hydration and structure formation in cement and chemical and mineral admixtures bared materials.

The general idea of the modelling is to predict early age strength and static modulus of elasticity growth of ordinary and high strength concretes.

Based on the porosity factor, a simulation model of hardening concrete was used for the prediction of concrete strength. Simulation models for development of temperature distributions and development of mechanical properties are used for the thermal stresses analysis method of the investigated large concrete columns. The **HPC** and **PC** columns had comparable maximum exertion values. The difference concerned only the time needed to achieve the maximum exertion values, which was 17 hours for **PC** and 32 hours for **HPC**. Initial temperatures of **PC** and **HPC** mixes were 24°C and 27°C, respectively. Influence of relaxation on concrete exertion during the hardening process was minimal. No cracking was observed on the surface of the two investigated columns.

## References to chapter 3

[3.1] Branco F. A., Mendes P. A., Mirambel E.: Heat of hydratation effects in concrete structures. ACI Materials Journal, v. 89, 2/1992.

[3.2] Eierle B., Schikora K.: Badensplatten unter frühem Temperaturzwang – Rechenmodelle und Tragverhalten. Bauingenieur, 75, 200, p. 671-678.

[3.3] Ekerfors K., Jonasson I. E., Emborg M.: Behaviour of Young HSC. Utylization of HSC. 20÷24 June1993, Lillehamer, Norway, v. 20.

[3.4] Emborg M.: Thermal Stresses in Concrete Structures of Early Ages, Lulea University of Technology,1989: 73 D.

[3.5] Godycki-Ćwirko T.: Concrete mechanics. Arkady, Warszawa 1982 [in Polish].

[3.6] Hirschfeld K.: Die Temperaturvertailung im Beton. Springer Verlag, Berlin Göttingen-Heidelberg 1948.

[3.7]   Kiernożycki W.: Massive concrete structures. Polski Cement, Kraków 2003 [in Polish].

[3.8]   Mikoś J.: Strength of concrete in the porosity factor function. Archives of Civil Engineering Vol. XXX 2/85.

[3.9]   Rüsch H., Jungwirth D.: Stahlbeton-Spannbeton. Band 2. Berücksichtigung der Einflüsse von Kriechen und Schwinden auf das Verhalten der Tragverte. Werner-Verlag GmbH, Düsseldorf 1976.

[3.10] Rostasy F. S., Henning W.: Zwang in Stahlbetonwän auf Fundamenten. Beton-und Stahlbetonbau, 8, 9, 1989, p. 208-214, 232-237.

[3.11] Simons   H.   J.:   Betonierabschnitte   von   Stahlbetonboden-platen   ohne Mindertbewehrung. Beton-und Stahlbetonbau, 94, 6, 1999, p. 254-258.

[3.12] Spriggs R. M., Vasilos T.: Effect of grain sizebond strenght of alumine and magnesia. J.AM. Ceram. Soc. 46, 224, 1963.

[3.13] Timoshenko S., Goodier J. N.: Theory of elasticity. Second edition. Mc Graw-Hill Book Company. Inc. New York, Toronto, London 1951.

[3.14] Van Breugel K.: Simulation model for development properties of early age concrete. Symposium RILEM, Ghent 07.1999.

[3.15] Weshe K.: Baustoffe für tragende Bauteile. Wiesbaden, Bauverlag, 1993.

[3.16] Witakowski P.: Thermodynamic theory of maturing. Application to massive concrete structures. Politechnika Krakowska, IL z.70, Zeszyt Naukowy 1, Kraków 1998 [in Polish].

[3.17] Eurocode 2: Design of concrete structures. Part 1: General rules and rules for buildings. ITB, Warsaw 1993 [in Polish].

# THE EFFECTS OF CEMENT MATERIALS HARDENING PROCESSES IN MASSIVE STRUCTURES

## Summary

In this publication theoretical and experimental studies of development of properties of early age concrete are presented. The use of high-performance concrete's means higher content of binders and lower water to binder ratios in comparison with the use of plain concrete. In this publication strength growth of seven concrete mixes with water to binder ratios between 0.52 and 0.32 are presented. A simulation model of early age concrete is presented for the prediction of development of the degree of hydration and structure formation in cement and chemical and mineral admixtures bared materials. Mathematical model showing the dependence between concrete compressive strength and porosity coefficient of its structure is presented as well. The general idea of modelling is to predict early age strength growth of Plain and High Performance Concrete's.

The setting and hardening of concrete is accompanied by no stationary, heterogeneous temperature and moisture distributions caused by the hydration heat development. In this publication theoretical and experimental studies of temperature distributions induced in large High Performance Concrete (**HPC**) and Plain Concrete (**PC** – no. 1A at the table 1.1) blocks during the hardening process are discussed. Two concrete mixes with water to binder ratios 0.52 (**PC**) and 0.32 (**HPC** – no. 6 at the table 1.1) are used in the experimental blocks to identify the temperature fields during its hardening. Composition and physical properties of the analyzed concretes mixes are shown in the chapter 1 of this publication.

Concrete structures treated by outer load are exposed to loads resultant from material properties and technology of structure. These loads are mainly caused by unsteady, coupled fields of temperature and humidity. They come into being during truss and hardening of concrete under the influence of action of inner heat sources and inner dry up as a result of hydration process in binder and also as a result of mass (humidity) and energy (heat) exchange with environment. These processes, causing irregular changes of volume in hardening concrete are often a reason of forming of first scratches and cracks in structure even in stage of its realization. Thermal and humidity distortions are essential in massive structures. Growth of massive structure temperature evokes distortion and stress in this structure. Maximal values of thermal stresses are in initial period of concrete truss and hardening. Cracks can be made in zones of tensioned concrete on the surface of solid. In the consequence of later, natural cooling of structure of the influence of plastic and rheological distortions and changes of material features in hardening concrete, there is a phenomenon of inversion in solid's distortion and it's causing tension of inner zones. In this publication thermal stresses of two experimental concrete blocks are presented. One of them was made with water to binder ratios 0.52 (**PC**) and 0.32 (**HPC**). Composition and physical properties of the analysed concretes mixes are shown in the chapter 1 of this publication but the temperature fields are shown in the chapter 2 of this publication.

## Note

I would like to inform you that this publication is a compilation of my previous works:

- J. Ślusarek: The correlation of structure porosity and compressive strength of hardening cement materials. ACEE, No. 3/2010, p. 85-92. The Silesian University of Technology, Gliwice 2010.
- J. Ślusarek: Hardening concrete effects in massive structures. EJPAU 2008 vol. 11 issue 2, p.1-19. Wroclaw University of Environmental and Life Sciences, Wroclaw 2008.

and has been published by permission of EJPAU and ACEE publishers.